mcommerce

The converging offline and online worlds:
A practical guide to exploiting the
fast growing mobile sales channel

9781 9082 9327 5999

RUPERT POTTER

CGW PUBLISHING

CGW
PUBLISHING

2013

mcommerce

The converging offline and online worlds: A practical guide to exploiting the fast-growing mobile sales channel

First Edition August 2013

ISBN 978-1-9082932-7-5

CGW Publishing 2013

CGW Publishing
B1502
PO Box 15113
Birmingham
B2 2NJ

www.cgwpublishing.com

mail@cgwpublishing.com

Cover images © CGW Publishing 2013 with thanks to S.Smith.

Contents

Index of Figures

Preface

Mobile commerce (also known as mcommerce) has been with us now for over a decade. Yet there are precious few examples of it making a real difference to the financial results of offline businesses.

Online businesses have been quick to open up and exploit the mobile channel by tailoring their websites for mobile browsers. But is there a difference between *ecommerce on a mobile phone* and *mcommerce*? And, if so, what is it? And how can it be exploited?

I work in mobile payments and that gives me a ringside seat to watch mcommerce develop. My colleagues and I frequently see companies struggling to understand what mcommerce can do for them. Or, once they've understood that, to work out how best to implement it. This leaves some organizations taking no action and others making basic errors. And all the time they see a third set growing a new revenue stream and improving customer satisfaction.

This book has been written to help bring structure and order to the thinking process of organizations considering how they might use mcommerce. Having read the book, the reader should be able to work out how mcommerce might help their organization, avoid some of the pitfalls when choosing between deployment options, and develop an implementation approach that balances risk with the ability to make good progress.

Acknowledgements

Firstly, I'd like to thank Keith Brown. Anyone knowing that Keith is my Managing Director might think thanking him first is the literary equivalent of deliberately losing to the boss at golf. Nothing could be further from the truth. It is no exaggeration to say that this book could not have been written without Keith's knowledge and support. I'd also like to thank Vicki Robbins for reviewing the draft and indeed everyone at Paythru who has contributed in some way.

I thank Christopher Greenaway and CGW Publishing for publishing the book and Alison Shakspeare for copy editing it.

Finally I'd like to thank Jacqui Barre for her help in reviewing the book as well as for giving me the space to write it.

Although all these people's input has been invaluable, I remain responsible for any errors or omissions in the text.

Rupert Potter

July 2013

Executive Summary

It is over a decade since the first smartphone was released and also since the first book on mobile commerce was published. For the whole of that period, the mobile phone industry has been predicting the imminent take off of mcommerce. That take off is now happening. Some companies are reporting strong growth in mcommerce revenues; mcommerce is credited with helping the turnaround of British retailer Argos; and data released by card schemes and other payment organizations show a dramatic increase in mcommerce sales. PayPal's figures show that it processed $2 billion of payments from mobile devices in 2010 and that the figure had risen to $20 billion in 2012.

Mcommerce was held back by both supply and demand side factors. Suppliers were unwilling to invest while it appeared that the technology was changing rapidly. Consumers faced with a choice of ecommerce on a PC or on a mobile phone often chose the greater ease of use of the PC.

Since Apple's iPhone was released, and a series of touchscreen-based smartphones followed, the technologies used in smartphones have been relatively stable. The basic elements required for mcommerce – screen, browser, app, camera, NFC, Wi-Fi, 3G – are all well established technically and in consumers' minds. This stability has given retailers (in the widest sense) the confidence to invest in mcommerce. The introduction of websites optimized for mobile devices has increased the ease of use for consumers. This ease of use, and the ability to use a mobile phone while out shopping, has driven the adoption of mcommerce by consumers.

One obvious difference between ecommerce and mcommerce is mobility, the ability for a consumer to make purchases almost anywhere. But is there a difference between *mcommerce* and *ecommerce on a mobile device*? A key difference is that mcommerce allows integration between the mobile device and the offline world in a way that benefits both consumers and retailers. Understanding and exploiting this difference is an important component of any mcommerce strategy.

A mobile device knows where it is, so retailers can provide offers to consumers as they approach their stores. The mobile device can start a conversation between the retailer and the consumer in other ways too, for example by tapping a phone against an NFC chip in a poster. Shopping can also be integrated. One of the case studies in this book shows how Mudo – a Turkish fashion and furnishing chain – allows customers to scan items on their phone as they put them in their basket and then allows them to pay using their mobile wallet. Proving that the payment has been made is another integration point with the retailer's offline environment. The whole transaction – both checkout and payment – is completed on the consumer's phone. This suits the consumer who values self-service as being more convenient and solves Mudo's problem of large queues building up at checkouts.

The mobile wallet also allows the ability to store offers, coupons and vouchers, and to collect and redeem loyalty points, providing a direct link between marketing and shopping activities.

Mobile devices are emerging as powerful advertising media. Consumers read text messages and push messages far more readily than they read emails, flyers and physical mail. The ability for retailers to match offers to consumer preferences at the time they are near their stores is about to revolutionize marketing. The case study in the chapter on mobile marketing shows fully 12% of push messages containing offers being converted into sales. The cost of acquiring a new customer is dramatically reduced allowing marketing budgets to deliver far more.

An important facet of mcommerce is the ability to pay, simply and by a variety of means. Many companies are now offering mobile wallets. Providers of ecommerce wallets are turning their attention to mcommerce and are now also providing mobile wallets. These wallets provide convenience for consumers and thereby help bring new customers to a retailer. Once a new customer arrives however, the retailer needs to take steps to convert that customer to its own wallet. Otherwise the wallet provider will remain the primary owner of the relationship with the customer. And many wallet providers have a vested interest in using the customer's data for their own purposes, which may include directing customers to the retailer's competitors.

So retailers need to provide wallet facilities both as a way of providing convenience to, and as a way of owning and maintaining their relationship with, their customers. As noted above, the wallet should be able to store offers, vouchers, coupons and loyalty points in addition to payment methods such as cash, cards and bank transfer details.

Mcommerce offers retailers the ability to make fast implementation progress in a relatively low risk way by using a combination of existing systems and processes and off the shelf products. For example, many retailers already offer a click and collect service. By tying that to a location-based marketing platform a retailer can send an offer, allow the customer to pay for it, and then have it ready for the customer to collect. Increasingly, the barriers to such implementations are internal and organizational rather than technical.

Mcommerce, especially when it is integrated with the offline environment, can be seen as extending the trend towards customer self-service. Where this is done well it is often welcomed by consumers. With mcommerce the consumer brings their own hardware with them. For payments, the mobile phone takes the place of a point of sale (PoS) device and, in the example of the bar code scanning app mentioned above, it also takes the place of the checkout.

So mcommerce offers:

- an enhanced ability to sell to consumers;
- enhanced convenience for consumers;
- a ready-made way for retailers to manage their relationship with consumers;
- reduced operating and infrastructure costs for suppliers; and
- relatively low cost running and implementation options.

With these advantages, it is easy to see why mcommerce revenues are accelerating quickly.

Reading Notes

This book has tried to shy away from the use of jargon. In some cases though, that is an unrealistic goal. A comprehensive glossary is provided at the back of the book.

The book is aimed primarily at businesses that sell goods or services to consumers from physical locations. The term *retailer* is used throughout this book to mean all such businesses. Indeed, the term can extend beyond business as many charities and public bodies can use these principles. Similarly terms such as *shop* and *store* are used to mean any physical location where consumers interact with retailers including theatres, restaurants, theme parks, sports stadia and many other examples.

Mobile Commerce Market

This chapter provides an introduction to the subject by defining and describing the subject that this book addresses: mcommerce. The chapter goes on to provide some context by describing the size of the market and summarizing forecasts for its likely growth over the short to medium term.

Definition of mcommerce

There are many definitions of mcommerce. In particular, vendors of different products and services tend to define mcommerce in terms of their own products and services. To avoid any confusion, this section describes what is meant by mcommerce throughout this book.

Clearly mcommerce has two aspects to it, mobile and commerce:

- Mobile means the ability for consumers to transact without being tied to a particular location. In practice this means being able to transact on a mobile phone or a similar device (for example a tablet computer, a PDA or an entertainment device such as an iPod Touch);

- Commerce is the ability for vendors to sell to purchasers, and for purchasers to buy from vendors (which are subtly different concepts). For the purposes of this book, purchasers will be consumers, as in individuals rather than other businesses.

Mcommerce is frequently defined to include mobile payments and mobile banking. In my definition a mobile payment constitutes one component of an mcommerce transaction but is not mobile commerce in its own right. Mobile banking, including other forms of financial services provided on mobile devices, is a separate activity from mcommerce. The purchase of digital content for mobile devices is excluded, as is mobile gambling.

So, for the purposes of this book at least, mcommerce encompasses the ability for vendors of physical goods (including tickets and even virtual tickets) to sell those goods to consumers via a mobile device.

Even within this definition, the phrase 'sell those goods to consumers via a mobile device' splits into:

- ecommerce, where the consumer is using a mobile device to order digital goods, or to order physical goods for later delivery;

- mcommerce, where the payment is genuinely mobile (so the consumer is not tied to a physical PoS device); and

- static payments on a mobile device where the mobile device is used by the retailer but the consumer must be at a precise location (for example, to make a payment by a card in a dongle-based reader attached to the retailer's mobile device).

So that leaves the question: is mcommerce different to ecommerce on a mobile device? mcommerce offers the opportunity to integrate the mobile marketing and shopping experiences into the retailer's offline environment. The ability to use the unique features of mobile devices when the consumer carries one with them in-store allows retailers to enhance the consumer's shopping experience and thereby increase consumer satisfaction and sales revenue. The retailers that have started this integration are the ones currently seeing revenue growth from the mobile channel.

This book primarily focuses on the second of the three bullet points above.

Background

Mobile Commerce

The current mcommerce market place has evolved from two distinct starting points:

- The first examples of mcommerce involved consumers sending a text message. The mobile operator would charge the consumer a premium rate for sending the text message and pass some of that charge on to the retailer. Early applications included purchases of content (such as ringtones, screen wallpapers and porn), as well as purchases from vending machines and of roadside parking. Payment by text also became a means for charities to raise funds.

- With the advent of smartphones, consumers started to use ecommerce websites on their phone web browsers. Initially these websites were difficult to use as they were designed for use on the larger screens found on PCs. However, as the number of consumers accessing ecommerce sites on their mobile phones increased, site owners began to redesign them to make it easier to navigate and read on the screen of a mobile device.

The original mechanism of paying by text has now been live for over 15 years, giving mcommerce a lengthy (relative to many technical innovations) track record. Building on that track record the prevalence of mcommerce is now increasing as the design of consumer user journeys improves and consumers, especially younger ones, actively look for interesting and different ways to use their phones.

Mobile Marketing

As well as mcommerce coming of age, mobile marketing is also starting to find a compelling offering. Initially, mobile marketing involved sending text (SMS) messages to consumers. Although these offered limited opportunities for describing or showing products, text messages had – and still have – an advantage over letters and emails in that they are almost all read. Better than that, they are mostly read very soon after they arrive. With the advent of colour screens on phones, MMS messages improved the presentation of the content of marketing messages. In 2012, Text Republic, a mobile marketing company, reported that almost 20% of links in SMS and MMS messages are followed, compared to just 4% in emails.

Once smartphones arrived and websites began to be optimized for smartphone screens, banner adverts were also optimized. Again, click through rates for adverts on mobile versions of websites are significantly higher than rates for sites viewed on desktops and laptops (0.24% as against 0.1%).

As phone manufacturers add more and more features to phones, marketing companies are finding ways to exploit those features to help companies interact with potential customers. Such features include push notifications in apps, bar code readers, QR codes, Bluetooth, location tracking, image- and sound-based searching. I will discuss these later in the book.

For an introduction though it is enough to recognize that companies are using these technologies to communicate with potential customers in ways that have never been possible before and which consumers find interesting and engaging.

Market Size

This section of the book aims to give readers an idea of the current size of the markets for mobile commerce and mobile marketing as well as an idea of how that might grow over the coming years.

The central tenet of the book is the convergence of mcommerce and the offline retail environment. There are still few solid examples where this has been successfully achieved long enough ago that growth rates can be measured and extrapolated. So, reliable predictions of the market size of mcommerce are hard to find.

There are many market analysts and market participants producing estimates and forecasts for the size of the ecommerce market, and specifically the ecommerce market on mobile devices. Some of this data – particularly from larger players and card schemes – can be of good quality. Other data may suffer from its publishers having a narrow viewpoint or a commercial interest in the impact of the data they are publishing. Even data from these sources can be hard to interpret because organizations use different definitions. For example, what is included under the heading *retail*? And some organizations include Russia or Turkey in Europe while others don't.

In order to try to dilute the inadequacies of any one survey, the data shown below in Figure 1 aggregate a number of recent surveys to produce an estimate of the market size and its possible growth.

UK Mobile Retail Spend Forecast

(£m aggregated from surveys and forecasts)

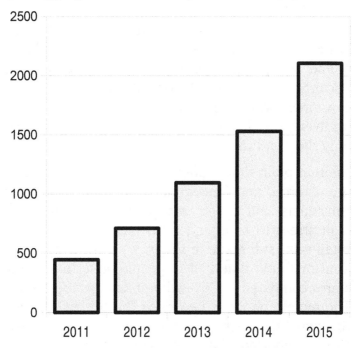

Figure 1: UK Mobile Retail Spend Forecast

As these data are aggregated there is a clear consensus among analysts and market participants that the market is growing rapidly and will continue to do so for several years to come.

Taking a wider perspective, surveys predict a similarly fast growth globally. A recent report[1] predicted that the number of consumers purchasing physical goods with their mobile phones would grow by 50% to 580 million in the next two years.

---

[1] Mobile Payments for Digital & Physical Goods: Opportunity Analysis 2012-2017, Juniper Research

A report by ABI Research predicted that in 2015, total worldwide mobile shopping revenue would reach $163 billion.

One interesting statistic is that consumers who switch to mcommerce also appear to embrace online shopping more quickly. Research by Forrester and PayPal (in 2011) found that consumers who shop on their mobile phones spend on average twice as much (£1,600) as those who only shop on PCs.

Another point that is often missed in the blizzard of ecommerce and mcommerce statistics is that consumers still prefer shopping in the mall and high street to shopping from home, with 90% of retail sales still made in shops. A survey in 2012 by YouGov found that 49% of people prefer to shop in stores compared to 29% who prefer to shop online. The mobile phone is becoming the bridge between the two. Also, 29% of consumers in the YouGov survey reported that they would shop on their phones (or shop on them more) if they could click and collect, which means buying on their phone and collecting from a local store. For many retailers, implementing a click and collect model is a relatively simple first step to implementing mcommerce.

It is not just offline retailers that are implementing click and collect models. Companies that have previously been ecommerce only, such as Amazon, are now providing collection points as an alternative to home or office delivery.

What's Working – Examples at Argos, Domino's Pizza and Starbucks

As well as understanding the overall market size, it is instructive to understand how successfully other retailers are using a mobile channel. This sub-section provides some examples of the impact that mobile commerce is having for some retailers.

Argos reported[2] that by the end of its 2012 financial year, sales via mobile had reached 6% of the total. For the financial year ending in 2013 mobile sales had grown by 117%. While these figures do not present a complete picture it is clear that the mobile channel is a significant and growing one for Argos. Indeed, in their plan to restore the company to sustainable growth the first item they talk about is: 'Leading in the market growth of digital commerce via online, mobile and tablet, while redefining multi-channel convenience for a digital age'. In other words, Argos not only sees the mobile channel as significant, it sees it as fundamental to the company's strategy. A large part of sales generated by the mobile channel is what Argos refers to as Online Check and Reserve, which is a first step towards the integration of the online and offline channels.

Domino's Pizza plc (the UK company) reported[3] that online sales for their year ending December 2012 were £268.6 million with almost 20% (£53

---

[2] Source: Home Retail end of year trading statements March 2012 and March 2013 and Investor Pack January 2013
[3] Source: Domino's Pizza Group plc Annual Results Announcement February 2013

million) coming from the mobile channel, with almost 9% from mobile devices. The percentage from mobile had almost doubled during the course of the year. Again, this shows how quickly the mobile channel can become important.

During its investors' conference call in April 2013, Starbucks reported that it is now taking four million payments each week in the US through its mobile apps. This accounts for 10% of its US revenue, which represents a doubling of volume since November 2012.

Summary

Market predictions, especially in a market's early stage of growth, can prove inaccurate. Mcommerce has been with us in different forms for over ten years now and its take off has been predicted before. There are now reasons for believing that mcommerce is a market that is establishing itself.

The technology has become relatively stable, giving companies greater confidence to invest. Some businesses are proving the value of that investment by demonstrating success in growing revenues from their mobile channel. Argos is an excellent case study of a business that has used ecommerce and mcommerce as a key plank in a turnaround. Finally, it is not only forecasts that are predicting growth in mcommerce sales, the measurement of those sales also records solid growth over the last few years.

Mobile Marketing

Introduction

This chapter considers how retailers are using mobile devices to communicate with their customers and potential customers. Despite having had the ability to advertise to consumers via their mobile phones for many years, the advertising industry is only just at the point of settling on techniques that are relevant to most advertisers and mobile phone users, and that are actually working.

Prior to the introduction of smartphones, advertisers were able to communicate by sending text messages. In situations where physical goods are not involved and the payment amount is small (for example, charitable donations or ringtone purchases) people can make purchases, or donations, by responding to these messages. Mobile Network Operators (MNOs) bill the consumer and pass the payment on to the payee. However, the limitations of such a journey mean that it will remain a small niche.

With the advent of smartphones, text message campaigns were able to include a website link for the consumer to click through to the substance of the advertiser's message. Text messages are a useful medium to the extent that they are almost always read as soon as they are received, so a high percentage of consumers see the message. Click through rates are high, often around 20% compared to 4%, or sometimes significantly less, for emails.

SMS-based marketing is analogous to deploying an email-based ecommerce marketing model on mobile devices. A simple message is sent to the consumer with the intention of enticing them to respond (by replying to the text or clicking through to a phone number or website) if they are interested enough in the initial message.

The web-based component of a lot of mobile marketing has essentially been to display ecommerce marketing pages on a mobile device. In order to make the most of mobile marketing, advertisers need to adapt to the unique characteristics of mobile devices and the ways in which consumers want to use them. These characteristics are:

- The device moves with the consumer (and can track those movements so that the consumer's location is known);
- The way that search is used; and
- The way that browsers are used.

The facts that the consumer now has a mobile phone with them almost all of the time; that the location of the device is known; that many consumers expect to use that device when they are shopping; and that they are happy to receive messages on it mean that these phones present an ideal medium for retail marketing.

Messages can be targeted at consumers when they are known to be in a location where they can respond to them. As the location is known, the offers can relate to circumstances – such as, say, the weather – at the location. This is known as

Proximity Marketing and is discussed in more detail later in this chapter.

One of the first applications for smartphones was the browser. This gave consumers access to the Internet while they were on the move. However, the screen size and lack of a keyboard (or the usability of keyboards when they are present) meant that consumers used browsers differently on mobile phones to the way they used them on desktop and laptop PCs.

Many website owners quickly realized this and implemented mobile versions with fewer and smaller pictures, less text, simpler user journeys and fewer, smaller adverts. But while those changes have retained traffic on mobile websites, they do not (and cannot) address the issue of how consumers use search engines.

Research into mobile browsing habits found that many users expect to find relevant content within 30 seconds of starting to look for it. When using search engines from a mobile device, they will visit significantly fewer sites than when running the same search from a PC. It seems that the decreased usability of mobile browsers means that users do not like mobile search and will settle for an adequate result rather than spend extra time searching for a better one.

A second finding of research into mobile browsing habits showed that a third of searches are for something local to the user, as opposed to around a fifth of PC-based searches looking for something local. This additional local component presents a marketing opportunity to offline businesses.

A third finding is that the time between running a search and taking action based on the findings is significantly shorter on a mobile device than on a PC.

These results are intuitively obvious. Clearly, using the ability to search when on the move will often be because the user wants something local and wants it while they are still in the locality. According to Google's (US) figures, 94% of smartphone users look for local information on their device and fully 90% take action as a result. And 65% of mobile users said they used their mobile device to find a business to make an in-store purchase.

This explains the importance of using the mobile channel effectively. The ability to reach consumers when they are nearby, in a way that increases the chance of them taking action, is something that retailers have been looking for for years.

Responding to the Search Issue

The fact that many users do not like search on a mobile browser has led advertisers to find different ways to attract consumers to their sites.

The first way is to ensure that advertising is displayed at the top of the search results and is most appealing to the user based on the search terms they have entered. The trouble is that everyone else is competing to do the same, making the cost and effort to succeed very high, which in turn reduces the returns.

Instead, there are two techniques that organizations are using to interact more directly with consumers: *push messages* and *mobile enabled advertising*. These are passive and active forms of advertising.

Push messages are sent (or pushed) to the consumer's device. They can be received by an app, or the operating system, and a summary of the message is displayed to the consumer. A ring tone or vibration will usually alert the consumer to the message's arrival. The consumer then clicks to open the message, reads it and responds. These messages are a passive form of marketing in the sense that the consumer takes no action, the messages just arrive.

Mobile enabled advertising means using bar codes, QR codes, NFC tags or SMS short codes (which are all explained later) in offline advertising such as leaflets, cards or posters. These are referred to as active because consumers need to take some action to scan them. Their ease of use means that they are much more likely to be actioned than, say, printing a web address, email address or phone number in the advert.

Both push messages and mobile enabled advertising can be forms of proximity marketing in which advertisers target consumers who are near a location where the product or service being advertised can be bought.

Proximity Marketing

Mobile Enabled Advertising

In this case, an advertiser places posters (or other material) close to their physical location. The advert encourages consumers to scan a QR code or tap an NFC tag in the advert. This takes the consumer to a website or app which offers an extension of the original adverting message. This will typically be an offer or voucher to tempt the consumer to the store but could include, for example, a video advert.

This is a simple but effective use case. Consumers scanning or tapping the poster are interested in the advert and so the resulting advertising message just needs to persuade its 'warm' audience to go to the right place.

Push Messages

There are many estimates of how much it costs a retailer to attract a new customer, covering a wide range, in other words there is little agreement on a figure. However, in a bell curve showing the customer acquisition costs in different sectors, many of those estimates fall in the range £15–£35. In a simple online advertising model, calculating a cost is far simpler. The cost per click for an advert tells the retailer how much it costs to attract a potential customer to their site. The figure divided by the conversion rate (i.e. the percentage of customers arriving at the site who go on to make a purchase) gives the cost per new customer.

A similar advertising model is now available for retailers attracting consumers to their store via mobile device adverts. In this model, a consumer is sent an offer, normally when they are in the vicinity of one of the retailer's stores. If the consumer is interested in accepting the offer they click on it and the retailer pays a price for that click. At the time of writing the price per click is in the region of 50p, similar to many ecommerce prices. Once they have clicked through to the offer, the consumer still needs take further action to complete their purchase. For example, they might pay online from their phone to reserve the item and then go to the store to collect it; or they may go to the store with their e-voucher and complete the purchase once they are there. Typically there is a further charge for this final step in redeeming the offer.

For this model to work, several elements must come together:

- The first concern is privacy. Clearly many consumers will react badly to unsolicited messages popping up on their phone, so push advertising is dependent on consumers opting in to the service;

- That in turn depends on consumers expecting to receive something of value in return for opting in. That could be content but is more likely to be some form of offer;

- Finally, the extent to which consumers are likely to put up with such advertising in the longer term depends not only on the value of offers but also on their attractiveness.

Combining the knowledge of the consumer's location, in order to send offers when they are in the right place, with knowledge of the consumer's habits and preferences, in order to send relevant offers, is known as context aware marketing.

Context Aware Marketing

As noted above, this form of marketing involves sending relevant messages to consumers' devices when they are in the right location to take action. This means being able to track both the consumer's location and their behaviour.

Tracking a consumer's browsing behaviour is especially important as it can provide a good indication of what that consumer is currently researching and is therefore likely to buy soon.

There is plenty of technology available on today's smartphones to allow location tracking:

- Tracking the masts from which a mobile device is receiving a signal and the strength of the signal from each mast;

- Tracking the mobile device's movement through a geo-location fence; that is when a consumer moves through a predefined circle of masts round a particular location;

- The device can connect to a Bluetooth transmitter with a known location;

- Similarly the device can connect to a Wi-Fi transmitter with a known location; or

- GPS (Global Positioning System).

From the consumer's viewpoint the key usability factor is battery life. A consumer is unlikely to leave their GPS switched on if it means their battery will not last throughout a day. The providers of location tracking systems will only succeed with systems that have a negligible effect on battery life.

Similarly, the technology already exists to track consumers' purchasing and browsing behaviour. These are typically behavioural algorithms, rule-based systems that process large quantities of data to find common patterns of behaviour which are then applied to data relating to each consumer to draw conclusions about what they might do next.

Context aware marketing is deployed via a marketing platform (or system) that combines the ability to track a consumer's location with making relevant offers to them via their mobile device.

Two distinct implementation models are emerging:

- one where the retailer implements such a platform to cover all their stores; and

- one where an area – such as a shopping centre, high street, train station, airport departure lounge or entire city – implements the platform to cover all (or many) of the retailers in that location.

In the first of these, the retailer has no competition for the offers it presents to consumers. It is purely a matter of finding an offer that the retailer believes is most likely to attract consumers and of setting the radius from a store within which the retailer believes the consumer is likely to take the trouble to visit. However, if several retailers in the same location have all implemented similar systems they run the risk that the consumer could be overwhelmed and ignore everybody's messages.

In the second model, with a platform covering all retailers in a location, a sponsor would implement and run the platform on behalf of the retailers. In a shopping centre that would probably be the centre's management company; in a city it might be a media company that already has advertising relationships with the local retailers.

In this environment there will be competition between retailers to send offers (or adverts) to consumers. This is because if all, or even many, of the retailers in a shopping centre try to send an offer to a consumer, the consumer will be overwhelmed and may well ignore them all, which would be bad for all the retailers. So a mechanism must be found to limit to a reasonable number the offers each consumer receives.

This competition could be resolved in many ways:

- Clout – the largest retailer in a shopping centre might demand the right to send the most messages. A department store taking up, say, 20% of the space might want to send 20% of the messages;

- Price – retailers might bid for the ability to send a message with the highest bidder's message being the one that is sent. The bids might be for specific days or even time of day (for example, cafés bid highest for lunchtimes). Or the bid might be for a specific consumer profile (with retailers bidding only to send messages to a particular demographic);

- Consumer preference – the provider of the platform serves offers that most fit a consumer's preferences. The preferences can be set either directly by a consumer when signing up to the service or can be set by tracking that consumer's behaviour over time within the shopping centre.

How this competition is finally resolved is likely to affect how successful such platforms are in the long term. In a shopping centre, both clout and price will be powerful incentives for the sponsor of the platform. However the consumer preference model is the one that is most likely to succeed in the long term, as the whole proposition is dependent on consumers getting value from it. It will be interesting to see whether sponsors of such platforms, and the retailers using them, are able to take this longer-term view.

What's Working – Promoqui

This chapter has described several different techniques for mobile marketing. A good way to bring this to life is through a case study of a relatively sophisticated implementation.

Promoqui is an Italian company with a background in producing advertising directories (similar to Yellow Pages). It has implemented a context aware marketing platform in Milan and Rome and uses a platform developed by Match2Blue.

Their implementation features:

- Location tracking;
- Consumer behaviour tracking; and
- Push messages.

Promoqui uses its relationships with retailers in Milan and Rome to allow them to join the Match2Blue platform, using it to manage details of different offers. Each retailer controls how much they are prepared to spend on click through, how many offers they'd like to send, the times of the day when they'd like to send them and the target demographic group for each offer.

Consumers are encouraged to sign up to receive offers as they move around their city. They can set preferences for whether they receive offers by push message or retrieve messages manually; how many messages they are prepared to receive and what types of offers they are most interested in.

Then, as they move around the city, the platform matches the consumer's location, preferences and previous habits to the offers that retailers have loaded into the platform. When the consumer is in range of a relevant offer it is sent to them. If consumers have elected not to receive push messages, they only see an offer in the app itself.

As of April 2013, Promoqui had signed up around 460,000 consumers, of which around 317,000 (69%) had elected to receive push messages.

On average, a consumer who elected to receive push messages received 51 per month. The click through rate was 35%, meaning that an average consumer was interested in 18 of the 51 offers they received - far higher than the click through rates typically experienced in ecommerce situations (which can often be under 1%). Of the 18 offers clicked, 6 were then redeemed. So, fully 12% of the offers sent turned into sales.

By comparison, retailers with an app that makes offers but without the context aware marketing and proximity marketing features, typically experience activation rates of 0.11 per consumer per month – one fiftieth of the rates experienced by Promoqui's retailers in Milan and Rome.

What's Working – Saatchi & Saatchi for Meat Pack

In another example, Saatchi & Saatchi developed a campaign called Hijack for their client Meat Pack. Meat Pack sells trainers and sneakers in a mall in Guatemala. Location tracking in the mall was able to identify when a Meat Pack customer entered a competitor's store in the mall, at which point they would receive a push message telling them they would receive a discount if they went to Meat Pack. The discount reduced by 1% for each additional second it took the consumer to reach the store.

This promotion 'hijacked' 600 customers from competitors' stores in its first week. Search online for 'Meat Pack Hijack' to find Saatchi's video summarizing the campaign.

Converging Trends

Merging offline and online worlds

One of the early behaviours that ecommerce stimulated was for consumers to do a large amount of online research before they went to a store to buy a product. They would often arrive at a retailer knowing what they wanted and the price that they could expect to pay for the product online. Consumers valued the service they received from a shop but they didn't want to pay a premium for it. John Lewis, for example, withdrew their 'never knowingly undersold' price match guarantee to combat this behaviour.

Mcommerce is stimulating almost the opposite behaviour. Consumers are going into a store, using the retailer's skills and knowledge to choose a product and then buying it on their mobile phone at the cheapest price they can find – often taking advantage of the warmth of the retailer's store while they do so. This is known as showrooming and is becoming a major threat, as well as a major irritation, to retailers.

However, the fact that potential consumers are in the store, interacting with merchandise and talking to the sales staff is also an opportunity for retailers. The trick is to make sure that consumers see the value in buying from the store while they are there: they will receive immediate gratification by leaving the store with the goods they want; they will receive better customer service if there is a problem with the goods after the purchase is completed; they could leave with a voucher offering a discount against a future purchase.

If the retailer is able to engage the consumer in the right way, they have the ability to strengthen their relationship with that consumer, for example, by collecting data from that customer for use in future marketing, or offering the consumer a discount on current or future purchases.

Developing a compelling user experience for consumers who want to use their mobile phone while in the retailer's store is half of what is needed to win the competition for mobile consumers. The other half is enticing them into the store in the first place.

This involves combining data about an individual consumer with knowledge of when they are near a store, at which point they can be offered a time limited offer for something that appeals to them to collect at the store. The purchase can be completed on a click and collect basis when they receive the offer or once they reach the store.

So the mobile channel offers retailers a compelling end-to-end proposition. They can:

- see when consumers are near their stores;
- make a relevant offer to them at that time;
- allow them to complete the purchase (either click and collect or in-store);
- collect data on consumer habits (either with or without a loyalty scheme); and
- allow the consumer to store an offer for use during a future visit.

The technology exists, and is well established, to deliver the entire proposition. Success depends on harnessing those technologies and using them creatively. The links between the online and offline worlds are still relatively unexplored. Both halves of this offline / online mobile proposition – attracting the consumer in the first place and tempting them to buy from the store – demand the effective combination of creative marketing skills and user journey design, underpinned by the effective application of technology.

This book goes on to explain the technology platforms available to retailers that support both halves of this proposition. The application of these technologies will depend on an individual retailer's strategy and position in the market. This book will give retailers the tools they need to exploit the merging of the online and offline worlds effectively.

Research into Offline – mcommerce Convergence

There is much research into how consumers are using their phones while physically shopping (as opposed to shopping online), and again, the data reported are variable and contradictory. The following research is from the 2013 Mobile Retail Summit and is consistent with much other recent research.

Smartphone Usage While Shopping

(% of smartphone users who have done each activity when shopping)

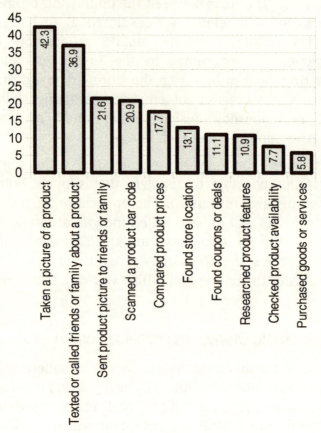

Figure 2: Smartphone Shopping Habits

These figures show both the opportunities and the threats faced by retailers. Consumers are using their smartphones to help them shop, looking for both stores and offers while they are out shopping. They are also comparing prices and buying online.

Mobile Wallets

At the time of writing, there is a huge amount of discussion in both the payments and retail industries about ecommerce wallets and mobile wallets. This chapter describes the key components of a mobile wallet and also reviews the issues that retailers will need to consider before settling on a wallet implementation.

Several ecommerce wallets that have now been launched by major brands and are gaining good traction with consumers and merchants, including:

- Google Wallet;

- Mastercard MasterPass;

- PayPal;

- V.me by Visa (at the time of writing this is launched in the US and about to launch in Europe).

These wallets were all designed for ecommerce applications. Their first implementations on mobile are also primarily ecommerce applications (in order to make payments to ecommerce sites optimized for mobile devices), although they are all now considering how their wallets can be extended into mcommerce. Of these providers, PayPal appears to be making the fastest progress.

As well as these large brands, there are now dozens of companies offering some form of wallet. Some of these can genuinely claim to be wallets, some are ideas capable of growing into wallets and some are aligning a narrow set of features with the industry hype.

Recent research[4] predicts that in-store payments from mobile wallets will reach €45 billion by 2017 in Europe. This compares to an estimated spend of €100 million in 2012. Although this is a vast absolute amount, it represents 1.6% of forecast card payments in 2017. In the US the report predicts in-store mobile wallet payments will increase from $500 million in 2012 to $44 billion in 2017.

Benefits of a Mobile Wallet

Provision of a wallet offers the merchant several benefits. These are covered in more detail in the text below but it is worth summarizing them:

- Consumer convenience – for the retailer's highest transaction volume customers the ability to store everything needed to transact with the retailer in one place makes their relationship with that retailer easier to manage. This in turn will enhance loyalty, or stickiness;

- Protection of customer data – data concerning the retailer's customers are protected from third parties who might want to use it either for their own purposes or to aggregate and sell on. Much of this data would often end up in the hands of organizations wanting to sell competing products;

---

[4] Berg Insight June 2013

- Reduced interchange (and acquisition) fees – used correctly, a wallet can help reduce interchange fees by reducing the number of payment card transactions subject to those fees. In practice, this will usually mean not allowing the use of credit cards either in the wallet or to fund the wallet.

Features of a Mobile Wallet

This section describes the features that might be found in a mobile wallet. Clearly some retailers will not want to implement the full set of features, however, to be worthy of the name, a wallet should be capable of providing the vast majority of features. A good analogy when thinking about a mobile wallet is a physical purse or wallet.

Cash

The first category is the storage of a cash balance. The obvious benefit of holding the consumer's cash in a wallet is that it supports the business' cash flow, in that it is being used to fund the business' working capital requirement until it is spent. There are other benefits:

- Once the cash is in the wallet, there are no interchange fees to pay for transactions, so thoughtful use of cash balances can reduce the retailer's overall transaction costs; and

- Cash transactions in a store are entirely anonymous, leaving the retailer with no customer data (unless they have used a loyalty card). A customer using a cash value from their wallet is generating useful data for the retailer about their shopping habits.

Consumers should be able to load cash into their wallets via a number of sources. Firstly, cash could be transferred from credit, debit or prepaid cards. The advantage of doing this is that the transaction-based element of any payment fees occurs once but supports several sales. This could be via an automated top-up, where there is a transfer from the card to the account balance every time the balance drops below a certain level. The other benefit is that data on the subsequent sale are hidden from the organizations in the payment infrastructure (and so the merchant's data are protected). However, card schemes are now increasing their charges for such transactions as the unavailability of this data reduces the effectiveness of their fraud scoring models and thus increases the risk of fraud.

Secondly, cash can be added via cash handling services (such as Paypoint and Ukash). The consumer pays money to a retailer and receives a receipt with a reference number in return. By linking the wallet to the appropriate provider, the consumer then types in the reference number to transfer the amount to their wallet.

Thirdly, by bank transfer. These transfers can be initiated by the consumer from their bank or from their wallet once the details are set up at the bank.

Fourthly, by transfer from the retailer's loyalty scheme. The links between loyalty schemes and wallets are discussed below but the ability to turn loyalty scheme points into a cash value in the wallets is likely to be attractive to consumers (even if they receive no extra purchasing value as a result, surveys show that cash is often valued more highly than loyalty points).

Finally, the wallet provider might also want to allow transfers into the wallet from other digital payment services.

Non-cash Values

As well as storing cash, the wallet provider should allow a wallet to store non-cash values. In some cases there might be no practical difference between a non-cash value and actual cash. For example if a consumer adds a £50 voucher to a wallet, is there any difference between storing the voucher and adding £50 to the cash balance? The types of non-cash value are summarized below.

- Vouchers – issued by the retailer should be able to be added to the wallet so that the voucher's value is safely stored and the original voucher can be thrown away. As noted above either the voucher itself can be stored or its value could be added to the cash balance in the wallet.

- Gift cards and certificates – if a consumer is given a value on a gift card or certificate, it should be possible to store that in the wallet. Again, this might be by storing the card or certificate or by increasing the cash balance.

- Loyalty values – as loyalty points are accumulated these should also be stored in the wallet.

- Conditional basket offers – there are many types of conditional offers but all could be stored in the wallet. A conditional basket offer refers to an offer for buying things in a particular transaction (such as 20% off the matching jacket if you buy a particular skirt).

- Repeat transaction offers – or a conditional offer might relate to more than one transaction (such as spend over £40 twice in May and get £20 off a transaction in June). The wallet should be able to remind consumers of these offers while they are shopping, and also remind consumers when an offer is approaching the end of its life.

Card Storage

The consumer should be able to store their credit, debit and prepaid cards in the wallet, this is a version of storing a cash balance. This ability would normally extend to the major card schemes. Typically, a consumer will be able to store more than one card in their wallet. An exception might be if a retailer issues their own credit card, then they might want to restrict the wallet to that card. However, most retailers will prefer to encourage a wide base of consumers to take up the wallet and so will want to avoid unnecessary barriers.

The consumer should also be able to link their wallet to their loyalty scheme membership. That way, points are automatically accumulated without consumers needing to carry physical cards with them.

There are two types of loyalty scheme. For an in-house scheme the wallet itself should be capable of storing the accumulated points. For a scheme that operates across a number of retailers (for example, Air Miles or the UK's Nectar schemes) the wallet should be able to report purchases back to the appropriate scheme via the retailer. So the consumer's membership number must be held and reported with each transaction.

Identity Card Storage

There are many other pieces of data that consumers routinely store in their physical wallets and purses. And there are ideas to allow their storage in electronic wallets. Driving licences and gym membership cards are good examples. They have little relationship to mcommerce (they could possibly be used as proof of age) and privacy issues mean that it is likely to be some time before their storage in digital wallets becomes routine.

Consumer Functionality

As well as storing cash, cards and other values, a wallet must provide a minimum set of functions to a consumer. These are:

- Account management – the ability to change their personal details, such as address and phone number; to add and remove cards from the wallet; to amend card details when they change; to change bank account details for fund transfers; and to change their password;

- Review recent transactions – the ability to see what they have bought, where and when;
- Balance – the ability to see cash and non-cash balances and to see a statement for each balance;
- Manual top-up – to be able to add cash to the wallet from a card or bank account;
- Auto top-up settings – how much cash is added, what triggers an automatic top-up and how is it funded;
- Review offers – to be able to see any current offers and their expiry dates.

Merchant Functionality

Similarly, there is a minimum set of functions required for retailers to manage the wallets.

Firstly, the merchant needs the ability to process refunds. The wallet must be able to match the refunded transaction to its original payment source so that the payment is refunded properly.

Secondly, if the retailer has sub-merchants the wallet will need to support accurate accounting for the amounts due to the sub-merchants.

Which Type of Wallet?

There are different types of wallets available for mcommerce. Choosing the right one is a matter of correct strategy. This section considers the two fundamental choices that retailers have to make to settle on the type (or types) of wallet that they will implement. The first choice is whether to use one of the generic wallets that are available across many merchants or to have an own-branded wallet. The second choice is whether to use a wallet that is stored in the cloud on servers hosted by the wallet provider or one stored on each consumer's phone.

Multi-merchant vs. Own-branded

The first decision is whether a retailer should use a multi-merchant or 'proprietary' wallet ; well known examples are the wallets provided PayPal, Google, Visa and Mastercard) or implement their own-branded wallet (that is only available for consumers to use at that retailer). This is an important strategic decision and needs to be taken with a great deal of care. This section outlines the strengths and weaknesses of each.

Multi-merchant Wallet

A multi-merchant wallet allows consumers to use their wallet across a number of different merchants. These wallet providers sign up large numbers of consumers (either directly, as in the example of PayPal, or through partners, as in the example of V.me by Visa that allows Visa's card issuing banks to sign up consumers).

The advantage of accepting these wallets is convenience for consumers. These wallets bring with them consumers who do not have to sign up to a new service to trade with the merchant.

There are some disadvantages to set against this. Firstly, some of the functions required in an ideal wallet will not be available and so will still need to be implemented separately by the retailer. These include integration into the retailer's loyalty scheme and storage of vouchers, coupons and offers. These wallets are designed for use across ecommerce merchant sites. Their full integration into mcommerce in the offline retail environment may prove difficult.

Secondly, the retailer will rarely get the cash flow benefit of any prepayments into the wallet since, typically, the wallet provider will keep them.

Thirdly, payment schemes increase the costs of cash transactions from these wallets (both loading cash into the wallets from cards and then making the payment to the retailers from the cash balance). This recognizes both the costs of maintaining the payment infrastructure and also the increased fraud risk associated with cash.

Fourthly, the wallet will require an additional login step. Even if the consumer has already logged in to the retailer's mcommerce site there must be an additional login step to authenticate the access to the multi-merchant wallet. This additional step in the user journey will increase the difficulty of completing the transaction and, in some situations at least, will increase the dropout rate.

However, perhaps the biggest risk facing the retailer is ownership of the customer data. For many years retailers have wrestled with the question of disintermediation; to what extent should intermediates be removed from the supply chain? Retailers wanted to disintermediate to give themselves greater control over the product and improve margins. Now however, wallet providers are threatening to become intermediaries of a different kind.

A wallet provider will accumulate data from all the transactions that pass through the wallet. Once a consumer starts to use the wallet in more than one retailer, the wallet provider will know more about the consumer's behaviour than each retailer does. The threat to the retailer is obvious; that the wallet provider will hold enough data about the consumer to sell competing products or services, or to allow a third party to sell to the consumer.

It is this ability to ensure that the relationship with the customer remains disintermediated, that makes the choice of wallet type a strategic one.

Own-branded Wallet

The pros and cons for a retailer of using an own-branded wallet are mostly the opposite of those listed above.

The disadvantage of an own-branded wallet is that, with many retailers offering them, consumers are only likely to sign up to a small proportion of those available. However, the ones that do sign up to an individual retailer's wallet are likely to be regular customers and consequently the most valuable customers. Other customers will need to be incentivized to sign up.

An own-branded wallet is likely to give retailers greater flexibility for its implementation and integration than a multi-merchant wallet. Retailers wanting to implement an own-branded wallet will have the choice of building their own, or implementing and customizing a 'white labelled' wallet (one where the functionality is built by the provider and customized for each retailer's implementation). In either case, the flexibility available, at least in the short term, will exceed that of multi-merchant wallets. This is because of the additional complexity of having to store and process loyalty scheme details, vouchers, coupons and offers across many retailers. Multi-merchant wallets will possibly offer this functionality in time but when, and to what extent, is not yet clear.

As noted above, any cash stored in an own-branded wallet will support the retailer's working capital. A question that often comes up is whether the ability to store a cash balance in a wallet is an activity that falls under the Financial Services Authority or European Union e-money regulations. At the time of writing it does not. Nor are there any formal proposals that it should be covered in the near future. The reason is that an own-branded wallet is a closed loop. That means that at the moment when the cash is stored, its end destination is known. Storing cash in an own-branded wallet is essentially the same as buying a gift voucher. Similarly, many restricted loop wallets (ones which can be used at a small predefined group of retailers) can store cash values without falling under the e-money regulations.

Finally, an own-branded wallet means retaining control over the consumer's data and not allowing a third party into that all-important relationship with the customer.

Cloud vs. Phone

The second choice is where the wallet and the data in it are stored. Are they stored on the wallet provider's servers and accessed over the Internet, or are they stored on the phone itself and accessed through an app?

Cloud-based Wallets

In a cloud-based wallet, the data is held on servers maintained by the wallet provider. The wallet can be accessed from the phone, either over a website or through an app (and it can also be managed by any device using a browser over the Internet).

By definition, the use of cloud-based wallets requires a connection to the Internet at the time of payment. So, either a good mobile data signal or Wi-Fi is needed to allow connection to the wallet.

The wallet can be accessed either from other web-based applications or from mobile apps and so is suitable for both mcommerce and ecommerce applications. This ability for the wallet to manage payments, loyalty point collection, offers, coupons and gift vouchers across both online and offline sales channels is a key reason for choosing cloud-based wallets.

Phone-based Wallets

Mobile Network Operators (MNOs) and handset manufacturers have championed phone-based wallets. The MNOs are looking for ways to extend the revenue they earn and have identified mobile payments as such an opportunity. There have been various attempts to build wallets, sometimes by one company acting alone and sometimes by groups of MNOs acting together. Some observers feel that this is so important to the MNOs that they have made it difficult for some of the cloud-based wallet providers to make their wallets available via mobile devices.

A phone-based wallet stores card details, in encrypted form, on a chip on the phone or SIM card. Loyalty scheme details can also be stored. The consumer manages their wallet through an app on their phone. Payment is made by tapping the phone against an NFC reader at the PoS.

Many phone-based wallets do not yet offer the full range of wallet features listed earlier in this chapter. In particular, the cash handling features are currently weak so retailers wanting to use a wallet to reduce interchange fees are (at the time of writing at least) unable to do so with phone-based wallets. And the app on each phone needs to be kept up to date. Finally, although phone-based wallets can be integrated into the physical retail environment, they do not lend themselves to easy integration into the retailer's ecommerce channel. If consumers cannot use their wallet to pay for, and collect loyalty points on, their ecommerce transactions, the retailer's relationship with them will become fragmented and their ability to use consumer data will be damaged. Some phone-based wallets require the use of NFC SIM cards, which must be stored in the phone before the wallet can be used.

There are two major structural issues arising from keeping the data on the phone. The first is that the wallet provider must maintain separate versions of their apps for each operating system they support. To get good coverage of consumers, apps must work under iOS and Android as a minimum, and preferably Windows, Symbian and BB10 too. Making a total of at least five versions of the app that must be maintained. Secondly, consumers with two (or more) devices may face difficulties. Someone who has separate work and personal phones would not have access to offers on both of them.

At the same time there is evidence that consumers favour websites over apps (this topic is discussed in more detail in the Implementation Approach section). All these factors are question marks over the long-term likelihood that the main delivery mechanism for wallets will be phone-based.

So, Which to Choose?

Figure 3, below summarizes the main points of each wallet type. As a summary this is both a generalization and a snapshot at the time of writing.

Which type of wallet to choose is a decision for each individual retailer. Without any context it is impossible to make a recommendation.

For the multi-merchant versus own-branded wallet decision, 'which to choose?' might be the wrong question. For many retailers, the best plan might be to use both together.

The own-branded wallet would be used for, and by, the most loyal customers. It is these customers who have the most to gain by going to the trouble of signing up for a wallet. And the retailer has most to gain by not sharing their data about these customers with third parties.

The multi-merchant wallets, however, will make it easier for new, and irregular, customers to transact and should therefore improve conversion rates for these customers.

The retailer can then concentrate on improving the take-up of the own-branded wallets by making offers to the most loyal customers that have not yet taken up a wallet.

At the time of writing, the decision between cloud- and phone-based wallets is easier – especially in Europe which is behind the US in the introduction of phone-based wallets. Although such wallets exist, their main functionality is limited to being a link between an NFC chip in the mobile device and a credit or debit card. So the combination of the fuller range of functionality and the ability for the wallet to operate across mcommerce and ecommerce channels will make cloud-based wallets the better option.

Wallet Type Decision Matrix

	Cloud-based	Phone-based
Multi-merchant	Originated from ecommerce wallets Can bring large numbers of consumers Lose ownership of the relationship with the customer (and control over data)	Tend to be built by MNOs Many just link a chip in the phone to a credit card Few offer the full list of wallet features Lose ownership of the relationship with the customer (and control over data)
Own Branded	Loyal customers likely to sign up and use the wallet Offer full list of wallet features Covers ecommerce and mcommerce channels Retailer retains ownership of customer relationship and data	Not yet commonly available This proposition is unlikely to suit many retailers

Figure 3: Wallet Type Decision Matrix

Key Technology Summaries

This chapter reviews some the technologies that are most commonly associated with mcommerce. It is not technical, in the sense that it includes any technical detail. Rather, it assesses the strengths and weaknesses of the key technologies and identifies situations in which each technology might be most useful.

The first point to note is that the rate at which new technologies are being developed and integrated into mobile devices has slowed. All the key technologies that are currently being used for mcommerce have been deployed for long enough, and in enough devices, to be considered widespread and reliable.

NFC

NFC stands for near field communications. In an NFC interaction, two chips communicate with each other by radio waves. One of the chips might be passive, meaning that it is not powered but takes its power from the energy in the radio waves sent from the powered, or active, chip. Passive chips are also known as tags. Normally, the two chips need to be within a few centimetres of each other to be able to communicate successfully.

The NFC chips in mobile phones are active (as they are powered) and so can be used to communicate with each other, as well as to read passive chips.

Almost all smartphones now being shipped contain NFC chips. The notable exception is Apple's phones. At the time of writing Apple is thought to be developing and testing the technology but not to have made a decision on whether or when it will be deployed in their phones.

There are three uses emerging for NFC chips in mobile devices:

- as a payment method;

- to identify a wallet; and

- to initiate a conversation between a retailer and consumer.

Payment Method

Contactless credit and debit cards use a passive NFC chip to pass card data to a contactless reader at the PoS. In the UK a consumer can use a contactless card to make purchases of up to £20 without having to enter their PIN. If a consumer's wallet or purse contains only one contactless card, the wallet or purse can be held up to the reader so the retailer does not even need to see the card.

There is no reason why the chip containing the details has to sit in a card. It can be attached to anything that the consumer might have with them, such as a mobile phone or a key fob. Some credit card issuers will now send their customers a chip with glue on the back to stick onto anything they normally carry. Barclaycard (which offers these chips) created an April Fool's Day campaign, titled PayWag, allowing dogs to pay by a chip attached to their collars (search online for the video).

So, the NFC chip in a mobile phone can be linked to a credit card to allow the consumer to make small purchases without entering their PIN. This is not a mobile technology *per se*, in the sense that it can be used entirely independently of mobile phones. However, one possible deployment option is to link a card to an NFC chip in a mobile device.

The key limitation of NFC as a payment method is the limit on the size of the tap and go payment. For retailers with a high volume of small payments – coffee shops for example – contactless can shorten queue times as well as being convenient for the consumer. The current £20 limit means that many retailers will not be able to use contactless payments for most of their transactions. Although NFC chips can initiate larger payments, the user must enter their PIN for authentication so the convenience of contactless payments is lost.

Identify a Wallet

In this use case, when the consumer taps their phone against the chip, the phone will open their wallet. This can be used on entry to a store to show the consumer the balance of cash in the wallet, or any offers, coupons or vouchers available for shopping.

When used at the PoS, the offers can be automatically removed from the balance and the consumer can then choose which payment method they use to complete the transaction.

The subject of integration between the mobile device and the retailer's systems is covered in more detail in a later chapter.

The use of NFC tags as an entry point to a wallet is being widely discussed but there are few implemented examples as yet. In terms of the central thesis of this book – that retailers should integrate mcommerce into their offline environments – some form of integration between the PoS systems and mobile wallets is essential. NFC offers one possible route for that integration.

Initiating a Conversation

As noted above, NFC chips are now gaining significant traction in terms of the number of smartphones shipped with one installed. Although this means that many phones are capable of making NFC payments, or of accessing a wallet based on reading an NFC tag, consumer take-up is only just starting to pick up. Many consumers are likely to want to familiarise themselves with the technology through non-payment uses of the tags before moving on to payment applications. The non-payment applications of NFC chips tend to be marketing based.

An NFC tag is placed in a poster or another physical object. A symbol or message on the poster encourages consumers to tap their phone against the tag. The data on the tag then tell the phone what to do next, which is normally to go to a specific web page or launch a specific app.

The web page or app will then deliver a further advertising message to the consumer. This might be additional content, or might be an offer that the consumer can use. In proximity marketing, the offer might only be available for a short time.

Domino's Pizza for example is using exactly this technology in its Windows 8 food ordering app. The customer taps promotional material at the counter and is then able to collect offers and redeem vouchers.

Bar Codes

A bar code is a machine readable pattern with encoded data. There are many different standards for bar codes. An example of a bar code is shown in Figure 4, this is a one dimensional bar code because the data is read on only one axis, left to right.

Sample Bar Code

Figure 4: Bar Code Encoding the ISBN for this Book

Bar code scanners are commonly available as apps for smartphones, and bar codes are one of the image types read by Google Goggles.

Their relevance to mcommerce is that consumers have started to use them while showrooming. That is, consumers are finding goods they like on the shelves or racks in a store and then scanning the bar codes on their phones to look online for a better price.

Retailers are now starting to produce apps which allow consumers to read the bar codes on their phones while they shop, so that the mobile device is replacing the hand held scanners that some supermarkets have been using. The consumer arrives at the checkout with a basket of goods already scanned and totalled. It is a simple piece of integration to allow the mobile device to make the payment. This provides convenience to the consumer and allows the retailer to reduce both queuing time and costs.

QR Codes

QR code is an abbreviation of quick response code. It is a two-dimensional form of a bar code, read on both the left to right and up and down axes. Two-dimensional bar codes have the advantage of being able to encode much more data than one-dimensional codes. An example of a QR code is shown in Figure 5.

Sample QR Code

Figure 5: QR Code Encoding the URL for Wikipedia's Page on QR Codes

Although there are many standards for two-dimensional codes, the QR code has gained in popularity for two reasons. First, the specification is openly published and, although the format is patented, the patent is not enforced, so it can be freely used. Both QR code generators and QR code readers are widely available. Second, it has been taken up by Google, which in turn has given the marketing industry the confidence to use it.

QR codes can be used in the same way as NFC tags. Consumers scan them and are given a link to a website or an app to launch. They are a little less convenient for the user than an NFC tag since the QR code scanning app has to be launched manually. With an NFC tag, as long as NFC reading is switched on in the phone settings, just tapping the phone against the tag will launch the app or go to the web page.

Short Distance Wireless Technologies

Several wireless technologies have been adopted by mobile phone manufacturers, including ubiquitous technologies such as Bluetooth and Wi-Fi and proprietary technologies that various device manufacturers have tried to introduce to give their devices an edge in the market.

In terms of mcommerce these technologies can be used for:

- Proximity marketing – because of their short range, once a device is within that range its location is known;

- Serving offers – although there are several recorded examples of offers being served over Bluetooth, they have not yet had a major breakthrough in terms of volume. One example is the rollout of screens in convenience stores in Canada, showing adverts and recommending that mobile phone users enable Bluetooth to receive offers from major brands such as Red Bull and Coca Cola;

- Serving adverts – this has become popular especially over Wi-Fi. Mobile device users value a Wi-Fi connection for the Internet access it gives them, with good speed while protecting their data allowances. Many locations provide Wi-Fi to tempt consumers in and may show adverts as part of the Wi-Fi login process.

Dongle-based Card Readers

There are now several companies offering card readers which plug into a mobile device. The point of these is that they offer mobility to the merchant rather than to the consumer. The consumer still needs to be present at the device to hand over their card and enter their PIN, so from the consumer's perspective they are static payments on a mobile device, rather than mcommerce. These readers suit small businesses that move around, for example market stallholders or tradesmen. The question mark over these readers is whether consumers will trust them. Many consumers now have first- or second-hand experience of skimming (card details being stolen through PoS devices which have been tampered with) in reputable stores, which might make people suspicious of using such readers.

Authentication Technologies

Authentication is the process of confirming that a consumer is who they say they are. This is important in mcommerce because the payment is a fundamental part of the process. From a retailer's point of view, the idea of using mcommerce is to gain extra revenue, and that can only be achieved by a payment. If the consumer is not who they say they are then the payment might be fraudulent, leaving either the retailer or one of the other parties to the payment liable for the fraudulent amount.

Single factor authentication relies on something that the consumer has or on something the consumer knows. Two factor authentication normally relies on something that the consumer has *and* on something the consumer knows.

Examples of something the consumer might have are:

- a credit or debit card;
- a phone with a phone number that the retailer knows; and
- an authentication token.

In mcommerce applications it is typically one, or both, of the first two of these that are used. An interesting new area of authentication is the inclusion of biometric authentication in mobile apps. The technology is not yet working to the point of reliability required for authentication but, with the ever-increasing processing power in smartphones, it is likely to be just a matter of time before iris scanning, fingerprint scanning or voice recognition is trialled as a mobile authentication method.

Examples of something the consumer knows are:

- a username and password;
- a PIN; and
- the CVV (or CV2) number on a credit card (use of the CVV can also be taken as a demonstration that the consumer has that card to hand).

The level of authentication used depends on the nature or amount of the transaction. Most transactions might be authenticated by a single factor but large value transactions or wallet changes, such as changing an address or adding a new card, might require a second authentication point.

Entry of a PIN that is automatically checked against the copy held on a card's chip is now accepted as a common authentication method at PoS devices using cards as the payment method. Both cloud- and hardware-based mobile payment providers are working on ways of using a credit card's PIN to authenticate mobile transactions.

They key point for retailers building authentication into a purchase journey is to balance the protection offered against fraud with the disruption to that journey. The more difficult the journey becomes for the user, the less likely they are to complete the purchase.

One particularly disruptive form of authentication on a mobile payment journey is 3D Secure. This form of authentication – where the card issuer validates that the card user is genuine by asking them to enter a partial password – has worked effectively in reducing fraud in ecommerce transactions. It is not widely used in mcommerce because the pop-up screen for the password entry is handled clumsily on mobile devices. This in turn makes dropout rates very high, often over 40%.

Fraud Detection

Fraud detection systems can be implemented at various points in the payment infrastructure by the merchant, the acquirer (and / or the payment service provider in the case of ecommerce transactions), the card issuer or the card scheme. For a chip-and-pin PoS transaction (as in in-store rather than online) the liability for fraud sits with the card issuer or cardholder rather than the retailer. Most issuers rely on their own fraud detection systems to mitigate their risk. For ecommerce and mcommerce transactions the liability remains with the merchant, who must then implement their own fraud detection or pay for the PSP to run fraud checks on their behalf unless they pay for additional authentication checks (such as 3D Secure).

There are two basic types of fraud checking systems:

- Rules-based systems – rely on a series of rules to identify potentially fraudulent transactions. Basic rules might compare the goods delivery address to the card billing address; or they might block all transactions where the IP address of the user matches a block list.

- Learning-based systems – are able to improve their algorithms automatically by looking at patterns of data in transactions that are eventually known to be fraudulent. Learning-based systems are normally combined with rules-based systems to give the best possible protection. For learning-based systems to be effective they must see a high volume of transaction data; the more data they see, the more they have to learn from. Most of the systems implemented by issuers, card schemes and larger PSPs are learning-based because the high volume of data they process allows the algorithms to achieve a high level of accuracy.

Both rules- and learning-based systems typically separate transactions into three groups: those least likely to be fraudulent are allowed to complete; those most likely to be fraudulent are blocked; and the transactions in the group in between are usually referred to fraud analysts who conduct further manual tests to assess whether the transaction is fraudulent.

Implementation Guide

So far this book has outlined the main facets of mcommerce:

- consumers are rapidly adopting their phones as a way of buying goods and services;

- a key difference between mcommerce and ecommerce on a mobile is that the mobile experience is integrated into the retailer's physical environment in some way;

- the mobile device is an excellent marketing platform; and

- a wallet brings convenience to the consumer and allows the retailer both to manage its relationship with the consumer and to protect its consumer data from data aggregators.

However, while understanding these points is an essential prerequisite to developing an mcommerce strategy, the implementation itself also requires an understanding of the particular issues that can arise.

This section reviews some of the common issues that retailers face and some of the approaches available to solve them.

Model for Consumer Interaction

A useful model for considering the implementation of an mcommerce transaction is shown in Figure 6.

Integration Between Mobile and Physical

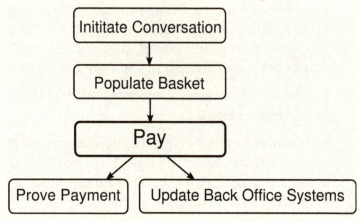

Figure 6: Simplified Model for Mobile to Physical Integration Points

Although this is a very simple model, it is useful to think about the interactions at each stage:

- What does the interaction between the device and consumer look like?

- What level of interaction is required between the device and the physical environment?

- How do the different components of the service interact with each other?

- What level of interaction is required between the mobile journey and the retailer's other systems?

Initiating the Conversation

This book has already covered many of the ways in which the conversation between consumer and retailer is initiated. Most conversations with the consumer will be initiated either by marketing activity or by the consumer being in the store and wanting to shop.

Examples of marketing activity initiating the conversation could be the retailer sending, or making, an offer by: push message; text message; email; including a tag or QR code on a poster for the consumer to tap or scan; including a link on a social media site or other ecommerce advertising

Any of these actions can initiate a shopping activity directly or it can take the consumer to a website or an app that displays further advertising material and then moves to a shopping activity.

The chapter on mobile marketing set out the benefits of communicating with a consumer at a moment when they can respond immediately, for example, using proximity marketing to communicate with them when they are close to the retailer's premises. This leads to the highest conversion rates. However, that does not mean it is the only activity worth pursuing. Sending an offer might initiate a conversation at some point in the future.

So many of the conversations at this stage require little or no systems integration work. The ones that do need integration work typically relate to offers. If the retailer is implementing a wallet, there may need to be systems integration or development work to allow offers to be loaded into a wallet. Generally, a wallet provider selling a white labelled wallet product should have built application programming interfaces (APIs) that allow offers to be loaded directly from the offers platform into a wallet, or as the result of a consumer receiving the offer and wanting to store it.

The other integrations will typically be very simple, using a tag, code or message to launch a shopping website or app.

Populating the Basket

In this step the word basket is used in its loosest sense. In a retail store, the word basket might be correct in its most literal sense, when the mcommerce basket might be populated at the same time as the physical basket. In other situations the mcommerce basket might refer to entering the amount to be paid. For example, at the end of a meal in a restaurant the waiter may hand over a bill. At that stage it is already known what has been bought (although the size of the tip might still need to be decided), so an mcommerce application here only has to register the size of the bill and allow the tip to be added.

These two very different situations (shopping or paying a restaurant bill) both require the mcommerce basket to be populated, but each requires a very different type and level of integration.

In the simplest integrations (for example, the restaurant situation described above), the retailer uses existing infrastructure and processes to determine what is being bought and how much must be paid. In other words, the whole of the populate basket stage is undertaken outside the mcommerce platform with only the result being passed to the mobile device.

The second way of the populating the basket is to use ecommerce infrastructure. In many cases this will involve re-using existing ecommerce sites. Earlier in this book we saw that Argos had made the mobile channel an important and fast-growing revenue stream. They did so by re-using their existing click and collect processes. Many attractions allow customers to buy tickets online before arrival. For customers without tickets, why not let them buy their tickets on their mobile phone rather than making them queue?

At the most integrated level, consumers would use their mobile device as the checkout till. Many stores, especially supermarkets, allow consumers to use a hand held scanner to scan products into their basket as they shop. The consumer then goes to a special checkout area to have the scanner read and to pay for the goods they have bought. For the consumer this avoids having to queue at the checkout; for the retailer this reduces the number of staff they need to run the checkouts. However, a significant amount of space is given over to storage racks for the scanners and to the checkout area. Not to mention the cost of all of the scanners. A simple mcommerce solution is to allow consumers to use their mobile devices instead of scanners.

A mobile app allows consumers to scan the bar codes of items as they place them into their trolleys. Fresh food items would be given bar codes when they are weighed (in the same way as with existing scanners). Once they have finished shopping, consumers check out on their phone. Any relevant offers in their wallet are applied before the payment is taken. Receipts could be printed in-store (with some till integration) or emailed. Wal-Mart is known to be trialling this technology.

What's Working – Mudo

Mudo is a retail chain based in Turkey with stores internationally. It has 130 stores as well as a thriving ecommerce presence and specializes in home furnishings and fashion. However, Mudo's success and popularity are leading to problems. Sometimes the checkout queues reach such lengths that customers cannot face the wait and many leave without completing their purchases. Mudo's response has been to implement an mcommerce solution, an evolution of an existing Smart Wallet app built by Shopamani.

Consumers download their app and register to use the service. This includes adding payment card details to their wallet, an integral part of the app. Mudo's popularity is such that in the four weeks prior to the launch of the service, before it was even announced, over 2,000 customers had downloaded the app.

The app works as described above. Shoppers scan the bar codes of items as they put them in their baskets. Once the consumer has finished shopping they pay for the goods using card details entered when they register for the service. Staff members at an express checkout area use tablets to see what customers have bought and to make spot checks to see that the app is being used correctly. The consumer's loyalty points are automatically accumulated with each purchase.

As well as the shopping experience this offers the consumer, and the reduction in dropout rates, Mudo also uses the app as a way of delivering loyalty offers to customers. A back-end system lets Mudo create and manage promotions. The promotions become an integral part of the consumers' interaction with the store.

So the app supports shopping, payment, loyalty points and offers; it provides convenience to Mudo's customers while they are shopping; it allows them to bypass the checkout queues; in turn that reduces the length of those queues, which allows Mudo to service more customers and increase their revenue without increasing prices. The investment is minimal as consumers are using their own mobile devices as both checkout and PoS device.

Making Payment

Once the basket is populated the next step is to pay for the goods or services. Payment will typically be through a wallet – either an own-branded wallet or a multi-merchant one – or by card. The integration points here are between the mobile shopping site and the wallet or payment process.

With a multi-merchant wallet the integration is normally fairly simple. The wallet provider will provide APIs for the retailer to use. The retailer tells the wallet provider how much is to be paid and hands over control of the process. The wallet provider then guides the consumer through their payment process and, once that process is complete, passes control back to the retailer along with a message saying whether or not the payment completed successfully. The basic integration process is the same for a shopping process built as an app or as a mobile website. From the consumer's point of view the process is a little more complex as it involves an additional login stage (to login to the wallet).

The integration between the mobile shopping site and an own-branded wallet is similar, but without the additional wallet login stage. The site uses APIs to hand over control and then to receive a message at the end about the success or failure status of the payment. With an own-branded wallet however, the consumer should feel that they are moving though an uninterrupted user journey. The branding remains the same, as does the look and feel of the site. If the retailer is providing the wallet using a white labelled solution (one bought from a supplier and then tailored with the retailer's own branding) it is important to ensure that there is not a separate login step. Consumers will be suspicious if it appears they have to login to the same site twice.

Finally, in the absence of a wallet, the retailer might use standard payment pages provided by the retailer themselves, a PSP, or a third party to complete the payment process. The payment pages will allow the consumer to enter their card and name and address details – much like on an ecommerce site – to complete the payment. Again, the consumer will view the entry of name and address details as an additional unnecessary step and this may lead to increased dropout rates. It is possible to build payment pages in such a way as to store the card details and name and address for future use. But if that is the intention, the retailer might as well implement a basic wallet in the first place.

Data Security

Data security is an important key concern in the payment process. The payment card industry has published a set of standards referred to as PCI-DSS (Payment Card Industry Data Security Standard), which cover how the payment card data must be protected when being processed and stored. The standard has different levels depending on the number of transactions being processed. At the higher levels, compliance must be audited annually. Companies handling smaller volumes can certify themselves using a self assessment questionnaire.

Compliance with PCI-DSS is important as businesses found to be non-compliant can face significant fines and be liable for fraud costs.

PCI-DSS is most easily managed through using a third party to host and initiate payments. A white label wallet provider (providing hosting as well as the software) and a multi-merchant wallet provider should maintain the necessary compliance. Similarly, a PSP, or other third party providing hosted payment pages, should provide compliance on behalf of the retailer. If the retailer is going to provide screens for card data entry, or provide infrastructure for data storage, they should approach their acquiring bank for guidance on the necessary compliance (this book deliberately does not include guidance on compliance as the compliance rules might change after publication).

Integration to the Payment Ecosystem

The infrastructure that processes payments on behalf of merchants is extremely complex. It has been built over decades and allows consumers with accounts at thousands of banks from all over the world to buy goods anywhere in the world. Fortunately access to this infrastructure is simple.

For offline (as in in-store) transactions, access is provided by point of sale (PoS) terminals supplied by the retailer's acquiring bank. For ecommerce transactions, access is provided by a payment service provider (PSP). The retailer should make sure that the PSP provides access to all of the payment types they might want to use. For example does a PSP provide access to specialized local payment types (such as iDEAL Payments in Holland) or to multi-merchant wallets (such as PayPal)?

Access to the payment infrastructure from an mcommerce application will also be via one of these two routes. The mcommerce app or website might access the payment ecosystem through a simple integration to a PSP, or via integration to the in-store PoS system.

The PoS system integration requires the mobile device to communicate directly with the PoS system. This is typically done by NFC communication with the PoS system interrogating the NFC chip in the device as if it were the chip in a contactless card.

Payments from Wallets

A wallet may allow several different payment types. When the time comes to make a payment, either the wallet software or the consumer (or a combination of both) must choose which payment type to use. A typical hierarchy would be:

1. offers and coupons (in turn prioritized by use by date)
2. vouchers
3. cash balance
4. debit card
5. credit card

The rationale for this hierarchy might be that consumers would expect to use offers and coupons first, so not doing so would create a customer satisfaction issue. Vouchers and cash (stored in a wallet) improve the retailer's working capital and using them to complete the transaction is free of charge. Then debit cards, which typically have the smallest charge, and finally credit cards, typically with the highest charge. Note that there may also be a hierarchy between different credit cards depending on the interchange rate the retailer pays for usage of each card type.

Proof of Payment

Once payment is complete the retailer needs confirmation to be able to release the goods and print the receipt. If payment integration was via the PoS itself, this step is not required (or rather, it happens automatically within the payment step). The integration options here are via the PoS terminal or by one or more API calls. Figure 7 shows these two integration routes.

Point of Sale to Mobile Integration Routes

The following routes refer to Figure 7 on the facing page.

Route 1: Direct from till system to mobile device (e.g. via a till scanner).

Route 2: From till system to mobile device via API calls between the till system server and the payment server.

Route 3: No direct integration between till system and phone. A tablet or PC connected to the payment server shows transaction data.

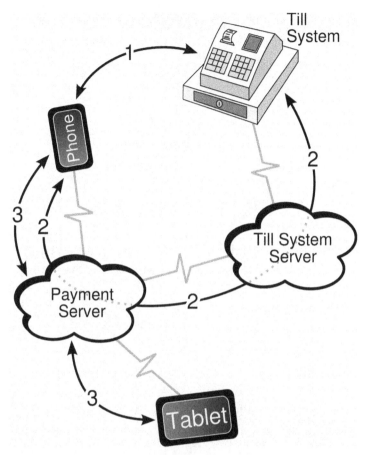

Figure 7: Summary of PoS to Mobile Integration Routes

A third option is to undertake this step with no integration by using the back office system of the PSP or wallet provider.

It would also be technically possible to provide an integration by a batch extract from the PSP or wallet provider's system. This option is not discussed as it applies to an ecommerce use case; it does not provide the real time information that a retailer requires for an mcommerce application.

Whether the integration is via the PoS system itself or indirectly via an API, the aim is to pass back to the PoS system information that the transaction has completed successfully and information about the payment (such as an authorization reference from a card issuer). Integration via an API tells the PoS system automatically, without the consumer or operator having to take any action, that the transaction is complete.

Integration directly with the PoS system requires that the system can read data from the mobile device. This can be by communicating with the NFC chip, but is more typically achieved by the mobile device producing a bar code (or QR code) that a scanner attached to the PoS system can read.

The final option is not to have direct integration at this stage. The person at the PoS is given access to the PSP or wallet provider's back office system, perhaps with a tablet locked down to access only that system. They can then use the system to check the validity of the payment.

What's Working – Natural History Museum

I recently went to see a photographic exhibition at the Natural History Museum. It was a last minute decision and so I had not bought tickets, nor had I realized that it was school half term week. When I arrived at the museum there was a large queue snaking through barriers to get to the front door. I asked a member of staff whether I needed to queue if I was only going to the photographic exhibition. He said I did unless I had already bought tickets. I asked whether buying them online was possible and whether that would let me bypass the queue? He said 'yes' to both.

So I found the exhibition website, found the tickets I wanted, entered my card details and bought the tickets. The website then gave me a message saying that there was nothing to print and that I only needed to show the number on the screen to gain entry. The member of staff let me through the fast entry. When I got to the exhibition entry, a member of staff took the number from my phone, looked for it on his back office system asked me my name and postcode to verify that I had indeed made the transaction, and let me in.

This is an ideal mcommerce use case. Buying the tickets online was faster for me and faster for the staff than if I'd queued and paid at the point of entry. No real time system integration was required since the staff verified my purchase by taking the booking number on my phone and checking it against the data recorded in their back office system.

Update Other Systems

During each of the stages above, the mcommerce system is generating data that can be used by the organization for other purposes. The data must be fed back to the relevant systems in the organization either through real time interfaces or as a batch extract.

The first major category of data is usage data. The mcommerce front end, whether it is an app or a browser, gathers data about how, when and where it is used. This data is required as it allows the optimization of the system, for example to minimize dropout rates or improve average transaction values. Many organizations are used to capturing and using this data already as it is similar in nature to the data captured by ecommerce applications.

The next category is basket data. This is captured so that stock management systems are kept up to date. Alongside customer data this is also used by marketing systems to track consumer behaviour.

The final category is payment data. Payment data must be fed back to the accounting systems to enable accurate financial control. It is also used by marketing systems to track consumer behaviour; to match offers and vouchers redeemed to those made and bought; and to manage loyalty schemes.

Much of this data is already captured in the stages outlined above. Clearly, where this is the case, a separate integration step is not required. Where a separate step is required the integration points are software-based – they do not rely on passing data over a wireless link to a PoS terminal.

Summary of Integration Options

The table in Figure 8 summarizes the mobile to physical integration options at each stage. This is not intended to be exhaustive but summarizes the options that will be most commonly considered.

Summary of Integration Options

Initiate Conversation
Manual initiation of app or website
NFC, QR code or SMS shortcode response
Receive offer (and respond immediately)
Receive offer (and store to click later)
Populate Basket
Pre-populate basket
Manual (select within website or app)
Scan bar code (or QR code)
Payment
PSP (potentially via a wallet)
PoS terminal (potentially via wallet)
Collection and redemption of loyalty points
Redemption of vouchers and coupons
Proof of payment
API integration to existing systems
PoS Integration (NFC, bar code)
No integration (access to PSP back office)
Other System Integration (all via API)
Usage data
Basket data
Payment data

Figure 8: Summary of Integration Options

User Interface Design

For many organizations, the fastest route to start selling on mobile devices has been to allow their ecommerce website to be available to consumers using browsers on mobile phones. While this is a quick route to mobile sales, it is not an effective one. Principles for the design of ecommerce sites are well established. However, they do not translate well to mcommerce. The two major differences between accessing sites by mobile device and a desktop or laptop device are screen size and bandwidth. The usability of the keyboard is a difference too, but many consumers are now proficient with smartphone keyboards so this does not need to be a major design factor in the user interface.

When using a mobile signal, download speeds vary enormously. This is a common experience for anyone who regularly uses a mobile browser while not connected to Wi-Fi. The download speed is affected by many factors, not just the type of signal. One variable is the number of devices connected to a particular mast and using the signal, download speeds at many city centre train stations are much faster in the middle of the day than during rush hour.

Mobile device users are not patient when it comes to accessing websites! A survey[5] of smartphone users said that 74% of users will abandon a site when a page takes over five seconds to load. So sites need to be designed so that pages load (or at

[5] Modapt and Morrissey Company August 2011

least the top of the screen that the consumer sees first) in a couple of seconds. Most sites will not achieve that over a 2G signal so the aim should be to load in under two seconds consistently on an EDGE (also known as 2.75G) signal.

The same survey reported that 60% of smartphone users said their biggest frustrations when mobile web browsing were sites that were difficult to navigate and information that was hard to read.

This sub-section gives some advice on user interface design for mcommerce applications.

Browser or App Based Interface?

The data on this decision are clear, users prefer using a browser to using a native app. This shows up consistently in surveys and in consumer behaviour. And the surveys show this is the case across all stages of the marketing and sales processes, including researching purchases, comparing prices, finding offers and vouchers, and completing purchases.

There are many stats showing how many apps are downloaded and used only once before being discarded. These statistics should be treated with a little caution since users can only try an app by downloading it. It is inevitable that some people will try an app and decide it isn't for them, so a fairly large dropout rate should be expected. Better measures are: 'which do consumers use when both are available?'; and 'how much revenue is generated by each?'.

Data on this are incomplete but that which exist suggest that websites outperform apps by over 2:1 on both measures. Anecdotally too, there is evidence of websites winning.

What's Working – Schuh

Schuh, a retailer that specializes in selling shoes, has both an app and a mobile-optimized website. They recently reported[6] that there is so little sales volume going through their app they are considering switching it off. They reached this decision when the revenue through their mobile site reached 15 times that through their app.

So, a retailer's first priority must be to develop a website that is optimized for mobile devices.

However, mobile apps cannot be discounted. Firstly because there are many examples of successful apps; both Starbucks and Dominos were cited earlier in the book as having successfully developed mcommerce revenue channels, both have very good and successful apps.

Secondly, the consumers that are most likely to use a retailer's app are the most loyal ones. The final reason why a retailer might develop and launch an app is that apps can do more; for example, integration with the phone's camera to read a bar code is much easier to achieve in an app. So, for mcommerce applications that require a high degree of integration into the physical environment, an app might be the only option.

---

[6] Mobile Retail Summit, April 2013, reported in Mobile Marketing Magazine

Do Not Force Registration

One of the benefits of mcommerce is that retailers can capture consumer data to allow marketing to the consumer in future. The temptation is to force this relationship by compelling new consumers to register with the retailer to use the site. In many cases this is counter-productive because the dropout rate at the 'register to use this site' stage is high and future opportunities to sell to the customer are lost, as well as the sale they were currently considering.

If the consumer likes the experience they will return and are likely to sign up voluntarily to receive the convenience of storing their card details and the benefits of receiving offers.

This point is basic customer focus, put the customer's needs first. One, unnamed, retailer is widely reported as having increased their sales by 45% (on an ecommerce site) by adding a 'checkout as a guest' button to their checkout process.

Minimize Data Entry Fields

This piece of advice helps with both the usability of the interface and the download speed.

The interface should only capture data that are absolutely necessary. There is no need to ask the consumer which search engine they used to find your site, or to ask for both home and mobile phone numbers. Use address lookup based on postcode and only ask for a full address if the postcode lookup fails. The design aim should be to complete any data entry process – and especially the checkout process – in as few keystrokes as possible, with as little scrolling backwards and forwards as possible and on as few pages as possible.

Use the Right Controls

What is the best way to allow a consumer to choose the number of a particular product they want to add to their basket? If the number that consumers normally choose is small (for example, when shopping in a fashion store), the best way is to pre-populate the field with '1' and use a '+' and '–' control to let the user increase or decrease the number. If the number of items covers a wide range (such as the number of shares bought in a transaction), the best way might be to let the customer type the number into the data entry field. Careful selection of controls makes the interface more intuitive and faster to use.

Pre-populate Fields Where Possible

Generally speaking, once a consumer has given data to a particular website, their expectation is that they won't have to enter it again on subsequent visits. There are some exceptions to this, like card data that they might not want stored and especially the CVV field which is not allowed to be stored. But for most frequently used fields, the retailer should be trying to use as much existing data as possible to save time and keystrokes to re-entry.

Checkout Navigation Tunnel

Once the consumer has selected 'checkout' they are normally committed to completing the purchase (a notable exception is that they might be trying to find out delivery charges if those charges are only added at the end). So, the theory is that once the checkout process has started the site should retain the consumer. That means providing as few routes out of it as possible. That, in turn, means providing as much help to consumers as possible within the process.

So, in the checkout processes there should be no:

- banner adverts;

- additional offers to increase the basket size; or

- links to other sites / pages (including social network links, which can be added to the completion page once the checkout process has finished).

Once a consumer is in the checkout process, give them confidence that the process is short by displaying a message such as 'page 2 of 3' or showing a progress bar.

Create a Secure Feeling

A recent survey[7] reported that 77% of UK consumers think that making payments on their mobile phones puts their money at risk. This is perception rather than fact as there is no evidence that consumers' funds are more at risk on a mobile

[7] British Retail Consortium / Google Online Monitor Q1 2013

device than on a PC (or indeed in a wallet or handbag). However, while the perception persists there are steps the retailer can take to improve security, and the perception of security.

Firstly, all payment activity should be over HTTP-Secure (or HTTPS) URL schemes.

Secondly, use text and iconography to reinforce key messages. Examples might be, 'your card details are not stored on the device', or a padlock symbol. While the payment is being made show a spinning timer and make sure it remains visible for long enough that the payment doesn't appear to be instant. Subconsciously, consumers expect security to add a little time to the payment process.

How Not To Do It

Google has produced a very good video on what not to do. It relates to ecommerce generally but equally applies to mcommerce. Search for 'Google analytics online checkout video'.

Implementation Approach

The first books on mobile commerce were published over ten years ago. As previously noted in this book, the technologies currently in use have not changed in the last few years. Yet, despite the stability of the technology environment and the length of time that retailers have had to exploit mcommerce, it is fair to say that no one sales and marketing approach has yet emerged to become ubiquitous.

That fact in turn calls for retailers to adopt an implementation approach which:

- addresses their own situation;
- allows flexibility to experiment and learn; and
- limits investment until a particular approach proves itself.

Such an approach is designed to allow retailers to make progress implementing mcommerce while limiting risk. The components of such an approach are summarized below.

Hands on Research

Before launching a new service, it is common practice to undertake consumer research. This is often done by questioning consumers in a store or by organizing focus groups. When launching mobile solutions, focus groups should be given hands on experience of using the solution in the environment where it would be deployed.

Building a proof of concept can be fast and cost effective, especially if it is supported by an mcommerce supplier who is keen to demonstrate the effectiveness of its solution.

So that might mean building a proof of concept and asking staff, loyal customers, or random consumers, to use it in a controlled environment. Once they have tried the solution they can provide structured feedback. The test can be run at a quiet time or even in the couple of hours after the business has finished trading for the day. The guinea pigs should be given access to the full system and tempted to make real purchases, either by offering them a discount on anything purchased, or being given prepaid cards.

Feedback on a full end-to-end test (from initiating the conversation, through the shopping and payment stages to the prove the payment stage) is valuable. It normally weakens the validity of a test if parts of the end-to-end process are excluded.

Buy Rather Than Build

Solution providers already offer all the software components described in this book. Retailers should seek providers that specialize in their field; provide pre-built software for the major components the retailer wants to implement; and provide easy integration to the retailer's existing systems.

In many cases, software is available on a pay-as-you-go basis or on a hosted basis. These options reduce implementation costs, timescales and risks and so offer an effective route to implementing proofs of concept.

Reuse Existing Processes

Both existing business processes and existing system components can and should be re-used.

Some retailers already allow consumers to buy something in the store and leave it for later collection so that they can carry on shopping. The process might already have been expanded to allow ecommerce click and collect transactions. Such a store has the core business processes and system components to allow consumers to scan the code for a heavy or bulky item, buy it and collect later.

A theme park might have scanners at its entrance gates to allow guests to have their ticket scanned to gain entry. A mobile application can produce a bar code in the same format so that guests can buy tickets on their phone (and so avoid queuing) and then have the code on their phone scanned to gain entry.

Many organizations considering mcommerce find that they already have existing processes that can be adapted and re-used.

Learn and Adjust

Implementation of an mcommerce solution should not be viewed as one monolithic project that has to be scoped and specified at the start and then implemented as a whole some time later. A large part of the reason for this is that there just isn't enough experience across a range of models to know which components of a solution will work best in which environments. So the approach needs to be one of trying something, adjusting to any lessons learned and then building on it.

If something doesn't work well when it is first implemented try to understand why, it might be an idea that is ahead of its time. In the earlier What's Working - Schuh sub-section we saw how they considered closing their app in favour of their mobile site. When their mobile site was launched in

2008 it generated 'no sales whatsoever'[8]. So it was switched off, then it was switched back on in 2011, by which time the take-up of mcommerce had advanced significantly. The mobile site now generates 46% of Schuh's ecommerce traffic. This shows the value of not giving up on a good idea just because it didn't work; and the value of changing approach once the reason for something not working is clear.

What's Working – M&S

In March 2013, The Independent published a story about Marks and Spencer trialling a mobile app in a small number of coffee shops in their stores. The app allows customers to order and pay using their mobile phones. This has all the hallmarks of the 'try and learn' approach outline above:

- a limited trial;
- with a simple integration;
- using existing business processes; and
- bought, rather than bespoke, software components.

Usability Testing

Testing for mcommerce projects should include usability testing on a range of devices, as well as the usual functional testing. This is true for websites as well as apps. The testing should include the major:

---

[8] Mobile Retail Summit, April 2013, reported in Mobile Marketing Magazine

- operating systems – at least Android, iOS, Windows, BB;

- manufacturers – Samsung, Apple, HTC, ZTE, Blackberry, Huawei, Nokia, LG Electronics, Sony, Motorola (the actual devices tested would depend on market share in the retailer's country and preferably among the retailer's target demographic);

- browsers – pre-installed browsers: Chrome, Firefox, Safari, Internet Explorer, Opera.

The requirement is to test as wide a combination of browsers, handset models and operating systems as possible to make sure that the pages render properly, and accept data properly, on as wide a range of devices as possible. This type of testing is easily outsourced if the amount of testing is too time consuming for an in-house team to attempt.

Choosing a Use Case

There are three components to choosing a scenario to use mcommerce.

The first is whether consumers are likely to use it. What is the reason that will tempt consumers to use their phones? Typical reasons include:

- access to offers;

- ability to avoid queues; and

- simplicity / convenience.

In some situations, access to offers might be replaced with access to content. For example, a customer buying tickets on their phone for a cinema might be offered access to see interviews with the film's stars.

'Ability to avoid queues' and 'convenience' can dovetail with the retailer's objectives. If a significant number of people use an mcommerce channel to avoid checkout queues it can allow a retailer to reduce checkout staffing levels, or even to reduce the number of checkouts and thereby increase the floor space available to display goods. The same is true of 'convenience'. Many consumers (although by no means all) see self-service as a form of convenience, and self-service can reduce the staff levels required.

The second component is clear benefits to the retailer. These benefits might include:

- revenue generation (or building market share);
- revenue protection (or protecting market share);
- collection of customer data;
- reduced cost of customer acquisition;
- reduced staffing costs;
- reduced transaction costs; or
- reduced infrastructure costs.

A customer who might walk into a store, pay and leave does so with no lasting relationship to the retailer. But, when that same customer pays for goods using their mobile device, or signs up for a mobile app, the retailer gains an opportunity to collect customer data for future marketing efforts. And that in turn gives the retailer a much greater chance of turning an occasional customer into a loyal repeat customer. For many retailers just the ability to collect customer data and establish a direct relationship with the customer is a valuable goal on its own.

In the chapter on mobile marketing we saw how retailers can use a context aware marketing platform provided by a third party (such as a shopping mall owner) to make offers to consumers. The cost of acquisition of such customers is dramatically lower than the cost of acquiring consumers through, for example, television advertising.

Over the past few decades retailers have gradually outsourced effort to their customers. For example, it used to be common for staff at a petrol station to fill customers' cars, or for people to go into a bank branch for their cash. Mcommerce offers the prospect of continuing this self-service trend, and the Mudo case study earlier in the book is a good example.

Mcommerce also gives retailers the potential to move tasks from hardware they would own to hardware that the consumer owns. In the Mudo example the consumers check out and pay by themselves and they do so on their own devices.

Their phone has taken the place of both the checkout terminal and the PoS device.

The third component is how well the implementation of such a use case fits the implementation guidelines outlined above.

An ideal use case is obviously one that has a compelling reason for consumers to use it; offers a strong business case to the retailer; and can be implemented in a low cost manner. Most organizations will be able to find use cases that score well across all three dimensions.

Conclusion

This book has described an mcommerce process starting with marketing, moving through shopping and paying and finally allowing the retailer to manage their relationship with their consumers. All of this done by using the consumer's mobile device.

The key differences between mcommerce and ecommerce are that the mobile device stays with the consumer most of the time. And it can be integrated into an offline environment to enhance the consumer's shopping experience.

This end-to-end, integrated use of the technology is relatively new. Some companies are now harnessing it to generate additional revenue, many others are wondering when and how to get started.

This book should help organizations to set off on the path of an effective mcommerce strategy.

Glossary

3G – third generation mobile communication standard; telecommunication standards for a mobile signal that provides a data signal fast enough to support many data requirements of mobile phone users (for example, browsing and streaming audio and video).

Acquirer – an organization, normally a financial institution, that provides merchants with the hardware and payment network connections required to accept card payments.

API – application programming interface, a way for a programmer to send data to, or retrieve data from, a system.

App (or Applet) – a program that runs on a mobile phone (or tablet).

Bar code – an optically read set of data. Initially bar codes comprised a series of black and white stripes of differing thicknesses. Consumers are most familiar with this type of bar code through its use in shops to identify specific products.

Bluetooth – a technology based on short wave radio, primarily used to allow mobile devices to communicate with accessories and other devices within a short distance.

CVV – card verification value; a number used on a credit card to increase confidence that the user of the card is its rightful owner.

Dongle – a piece of hardware that connects physically to the device (or indeed any computer) that allows it to perform an additional function.

GPS – global positioning system, based on signals from satellites and available on mobile devices to find their current location.

Interchange fees – the fees charged within the card payments infrastructure for processing card transactions.

mcommerce – abbreviation of mobile commerce.

MMS – multimedia messaging service, essentially a message to a mobile phone that contains images and / or video content, as well as text.

MNO – mobile network operator, a company that builds and operates a mobile network (or rents capacity from another company that does that) and sells contracts to consumers to access the network, for example Telefonica and Vodafone.

NFC – near field communications, a method for devices to communicate by bringing them close together.

PCI-DSS – Payment Card Industry Data Security Standard, which covers how payment card data must be protected when being processed and stored.

PDA – a personal digital assistant, essentially a palm top computer.

PIN – personal identification number that a consumer knows which allows them to identify themselves as the valid user of a particular credit card account.

Platform – in this book the term platform is used to mean all aspects of a system. So, for example, an offers platform is a system or group of systems that allows a merchant to manage offers.

PoS – point of sale; the point in a store where purchases are completed. PoS devices, PoS terminals and PoS systems refer to the hardware and software used to make / accept payments as well as to record the purchase and print receipts.

PSP – payment service provider; an organization providing electronic payment services to connect a merchant's systems (especially ecommerce websites) to the payment infrastructure.

QR code – quick response code, a matrix bar code. It holds more data than a standard bar code and is typically used to direct a mobile device to a particular web page. This example links to the author's employer's website:

Showrooming – the behaviour of choosing goods in a physical store and then buying them online via a mobile phone for the lowest price available.

SIM card – subscriber identity module card; a chip stored in the phone that is used by the phone to link to the appropriate number and subscriber account.

Smartphone – a mobile phone with a relatively high level of computing power offering users the ability to choose which applications they want to store and run on their phone.

SMS – short message service, commonly referred to as text messages.

Tag – a passive NFC chip, one which responds with some data when it receives power via a radio signal.

URL – stands for uniform resource locator and means a web address.

Use case – a particular situation or scenario in which a system is used; a description of the use to which the system is put.

User journey – a term used to mean the flow of screens a user sees, and data fields they enter, to complete a particular task.

Wi-Fi – a radio wave based technology primarily used to allow devices to access the Internet via access points positioned close to them.

About the Author

Rupert Potter has spent over thirty years working in IT and a significant portion of that time working on payment systems. He worked on the initial implementation of CHAPS and also on the UK's first debit card (Barclays Connect). His roles have included the development and implementation of IT Strategy and Business Strategy. Recently he worked as a consultant to Visa Europe on their V.me by Visa wallet.

He currently works for Paythru Limited, which specializes in providing payments for mcommerce solutions. Paythru's product allows merchants simple, ready to deploy, access to a wide range of payment types. These can be easily and quickly integrated into a mobile website or app.

He has an MBA from Henley Management College. His previous book, *Solving the Year 2000 Problem*, was published as part of the Financial Times Mastering Management series.

Paythru specializes in mobile payments. Neither Paythru nor the author have any commercial interest in seeing the mcommerce market evolve in the way described in this book. Paythru's commercial interest is in seeing the overall number of mobile payments increase rather than the growth of any one particular method. In fact, the author has used his independent vantage point in the mcommerce market to observe and describe where the market is moving.

The author can be contacted at:

Paythru Limited, www.paythru.com, 01494 415161

Index

S

T

V

W